TURTLE TALE

FRANK ASCH

THE DIAL PRESS · NEW YORK

The Dial Press
1 Dag Hammarskjold Plaza
New York, New York, 10017

Library of Congress Cataloging in Publication Data
Asch, Frank. Turtle tale.
Summary: A young turtle learns how to be a wise
old turtle, with a few hard knocks along the way.
[1. Turtles—Fiction] I. Title.
PZ7.A778Tu [E] 78-51328
ISBN 0-8037-8782-0
ISBN 0-8037-8783-9 (lib. bdg.)

To Dot and Ernie

One morning on the way to the pond
an apple fell on Turtle's head.

It hurt so much that Turtle pulled his
head inside his shell and made up
his mind to keep it there, thinking,
"That's what a wise turtle would do."

Inside his shell it was so dark
he couldn't see a thing.

But he could still smell his way
along the path to the pond.

On the way he bumped into an old friend.

He bumped into rocks,

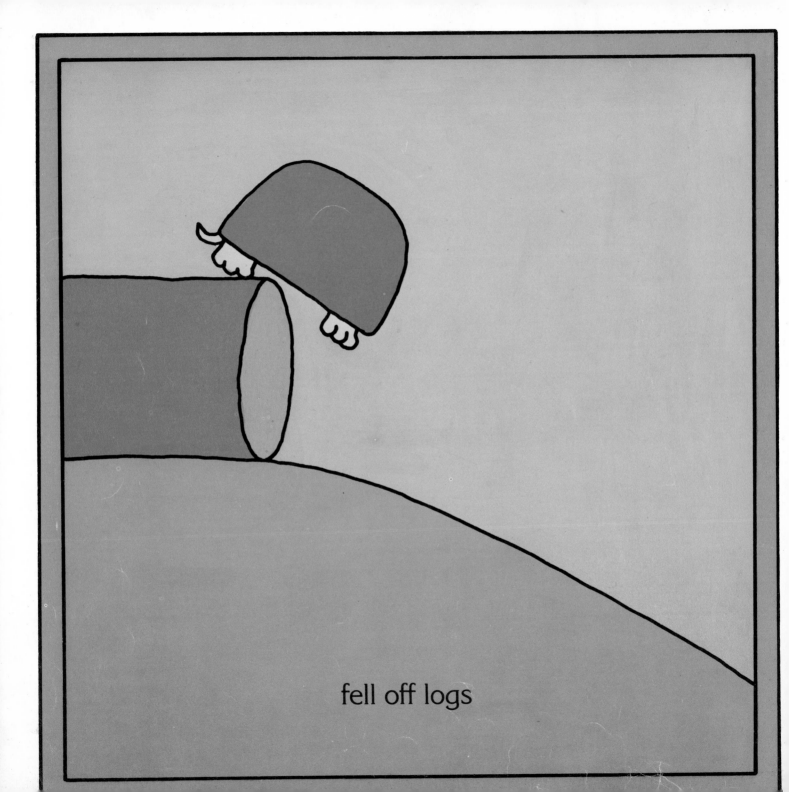

fell off logs

and tumbled down hills.

When Turtle got to the pond,
he could neither eat nor drink.

That night Turtle cried himself to sleep.

When he awoke the next morning he
stuck his head out and made up his
mind to keep it there, thinking,
"That's what a wise turtle would do."

By then he was very thirsty and
very hungry. He took a long drink

. . . and had a fish for breakfast.

After breakfast he climbed on top of his favorite rock and began to sun himself.

When the sun went in

and raindrops began to fall,

all the other animals took cover,

but not Turtle!

Fox too was running for cover when he saw Turtle sunning himself in the rain,

and decided to have him for lunch.

While Fox leaped through the air
straight for Turtle's head,

Turtle thought to himself, "But then again, maybe it's best if I keep my head out sometimes and sometimes pull it . . .

in."

Tucked safely inside his shell,
Turtle waited for Fox to go away.

When the rain stopped and the
sun came out again,

he finished sunning himself, thinking,

"That's what a wise turtle would do."